Animals in Forests

Julie Murray

Abdo
ANIMAL HABITATS
Kids

abdobooks.com

Published by Abdo Kids, a division of ABDO, P.O. Box 398166, Minneapolis, Minnesota 55439.
Copyright © 2021 by Abdo Consulting Group, Inc. International copyrights reserved in all countries.
No part of this book may be reproduced in any form without written permission from the publisher.
Abdo Kids Junior™ is a trademark and logo of Abdo Kids.

Printed in China

052020

092020

THIS BOOK CONTAINS
RECYCLED MATERIALS

Photo Credits: iStock, Shutterstock

Production Contributors: Teddy Borth, Jennie Forsberg, Grace Hansen

Design Contributors: Candice Keimig, Pakou Moua, Dorothy Toth

Library of Congress Control Number: 2019955549
Publisher's Cataloging-in-Publication Data

Names: Murray, Julie, author.

Title: Animals in forests / by Julie Murray

Description: Minneapolis, Minnesota : Abdo Kids, 2021 | Series: Animal habitats | Includes online resources
and index.

Identifiers: ISBN 9781098202095 (lib. bdg.) | ISBN 9781098203078 (ebook) | ISBN 9781098203566
(Read-to-Me ebook)

Subjects: LCSH: Animals--Habitations--Juvenile literature. | Habitat (Ecology)--Juvenile literature. | Forest
animals--Juvenile literature. | Forests and forestry--Juvenile literature. | Forest animals--Behavior—
Juvenile literature.

Classification: DDC 591.52--dc23

Table of Contents

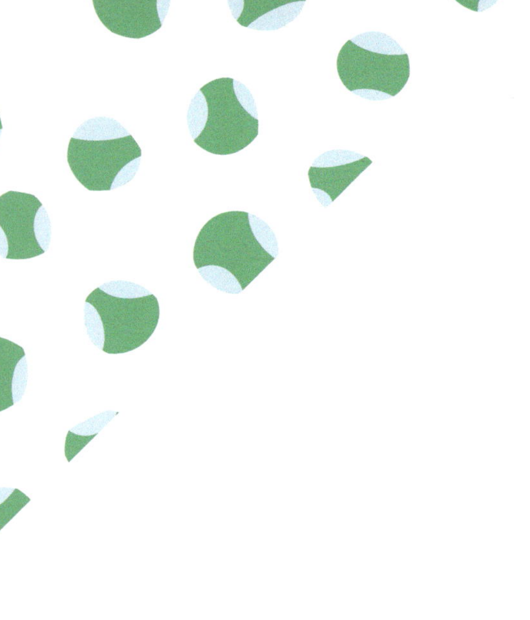

Animals in Forests

Many animals live in the forest.

Birds build nests high

in trees.

Rabbits dig **tunnels**. Many of them live together.

9

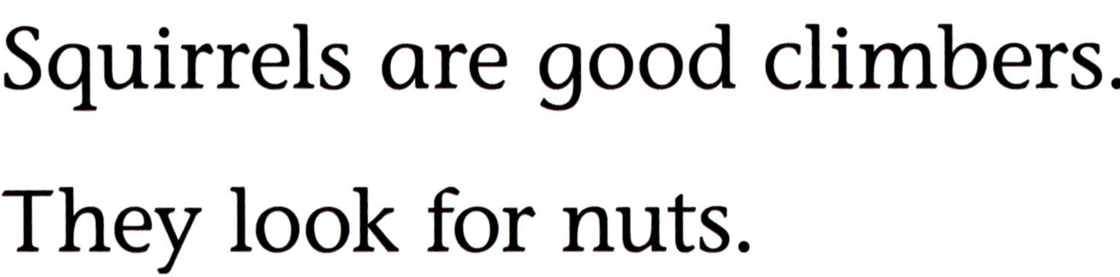

Squirrels are good climbers.

They look for nuts.

Foxes hunt mice. They hide
in the tall grass.

Deer like to eat berries.

Bears sleep in a **den**. They stay there most of the winter.

Owls hunt at night. They eat
small animals.

Moose are big. Their **antlers** can weigh 40 pounds (18 kg)!

21

More Animals in Forests

green tree python

raccoon

skunk

wolf

Glossary

antler

one of a pair of bony growths on the head of most kinds of deer.

den

the resting place of some large, wild animals.

tunnel

an underground hole of a burrowing animal.

Index

Abdo Kids ONLINE

FREE! ONLINE MULTIMEDIA RESOURCES

Visit **abdokids.com** to access crafts, games, videos, and more!